Cracking Neurosurgical Infections Vignettes
First Edition

Paul Edward Kaloostian MD, Sean William Kaloostian MD, Carolyn Louisa Kaloostian MD, MPH

Copyright 2012 by Paul Kaloostian MD

All rights reserved. No part of this publication may be reproduced, distributed, or transmitted in any form or by any means, including photocopying, recording, or other electronic or mechanical methods, without the prior written permission of the publisher, except in the case of brief quotations embodied in critical reviews and certain other noncommercial uses permitted by the copyright law. For permission requests, email to the publisher, addressed:

--Attention: Permissions Coordinator, at the address below:

paulkaloostian@hotmail.com

Printed in the United States of America

First Printing, 2012

ISBN: 978-1-105-98900-1

About the Authors:

Paul Kaloostian MD: Matriculated through Thomas Haider Accelerated Biomedical Sciences B.S./B.A./M.D. Program with Undergraduate work at University of California Riverside and Medical School at David Geffen School of Medicine at UCLA. Currently, he is Chief Resident at University of New Mexico Medical Center in the Department of Neurosurgery. He will be attending Johns Hopkins Medical Center for his Complex Spine and Spinal Oncology fellowship in 2012. He has a passion for treating the underserved and is fascinated by the immense cultural diversity in New Mexico. He is fluent in three languages: English, Spanish, and Armenian. He is an avid pianist and concert clarinetist and composes classical and Armenian Folk music.

Carolyn Kaloostian MD/MPH: Matriculated through Thomas Haider Accelerated Biomedical Sciences B.S./B.A./M.D. Program with Undergraduate work at University of California Riverside and Medical School at David Geffen School of Medicine at UCLA. Currently, she is chief resident in the Department of Family Medicine at University of Southern California Medical Center. She is currently obtaining her MPH degree at UCLA during her residency. She will attend UCLA Medical Center for her geriatrics fellowship in 2012. She has a passion for treating the underserved populations. She is an avid ballerina having performed in multiple national performances. She is fluent in three languages: English, Spanish, and Armenian.

Sean Kaloostian MD: Matriculated through Thomas Haider Accelerated Biomedical Sciences B.S./B.A./M.D. Program with Undergraduate work at University of California Riverside and Medical School at David Geffen School of Medicine at UCLA. He is currently a resident at University of California at Irvine specializing in Neurosurgery. Of note, he is a National Rhodes Scholar Finalist and Varsity Baseball player. He has also completed many marathons with competitive times. He has a passion for treating the underserved communities. He is fluent in three languages: English, Spanish, and Armenian.

Editors-in-Chief

Paul Kaloostian MD- Fellow and Instructor, Department of Neurosurgery, Spine and Spinal Oncology, Johns Hopkins Medical Center
Carolyn Kaloostian MD/MPH-Fellow, Department of Geriatrics, UCLA Medical Center
Sean Kaloostian MD-Resident, University of California at Irvine Department of Neurosurgery
William Kaloostian MD-Hospitalist, Internal Medicine Clinical Professor, Director of Short Stay Observation Unit, Kaiser Permanente Medical Center, Los Angeles, Ca.

Special Thanks/Dedications

This was a Herculean Task that could not have been possible without the efforts of many people. Special thanks to Eddie and Nanny for instilling with us the desire to teach and promote education, as well as a duty to give to others who are less fortunate. Thanks to Aida and Bill for their continued support and encouragement. They have dedicated their life to medicine and have always emphasized that knowledge is the key toward taking care of our patients and learning about oneself. We are endlessly appreciative of our many professors and teachers who have been unique role models in our lives. Finally, we thank our students and readers of this book and wish them a lifetime of happiness and education.

Preface

The medical literature is enormous. It is filled with information that is growing as you are reading this sentence. Having entered the realm of medical school and residency, it is very difficult to gather and master the information that is most critical. We have attempted to create a more concise and focused text that addresses the heart of the major issues that are not only commonly tested on Neurology and Neurosurgery Board exams but also encountered as one is taking care of sick patients. We have put much thought and effort into providing a text that can be used by pre-medical, medical students, nursing students, residents, as well as attending physicians and those in all scientific fields. Our goal is to provide an avenue of knowledge that can be used to heal that which is most important to us: Our Patients! Enjoy!

-Paul, Carolyn, and Sean

Note: All images are obtained from Authors (Kaloostian 2012).
Cover pic: Spectroscopy used to distinguish the molecular properties of various intracranial properties (Kaloostian 2011)

We are always looking to improve this book! Please email any questions, comments, or concerns regarding the book and improvements that can be made to this edition to the following email address: paulkaloostian@hotmail.com.

Table of Contents

I. Clinical Cases

INTRODUCTION:

The central nervous system is a fascinating anatomical specimen. The brain and spinal cord, and all related nervous tissue, are such delicate and critical structures that truly make us who we are as functioning human beings. Physicians and others in the health care team have the privilege of taking part in learning about the structure and function of the central and peripheral nervous system. We are given a large responsibility to care for patients who have unfortunately succumbed to illness in this particular area of the body. One such problem is infection of the central and peripheral nervous systems. There are a wide variety of infectious disease states that afflict this area and we have attempted to describe and teach many of these disease states in this textbook. We have written this textbook in the form of clinical vignettes of real-life patient cases that we have each taken care of during our training in medical school or residency/fellowship. We have learned so much from these patients and we wanted to share this teaching with the rest of the world. We hope that this textbook can enlighten readers of all specialties and fields about infectious diseases of the central nervous system and methods of diagnosis and management of these neurosurgically and neurologically impaired patients. Please enjoy!

Paul E. Kaloostian MD
Sean W. Kaloostian MD
Carolyn L. Kaloostian MD

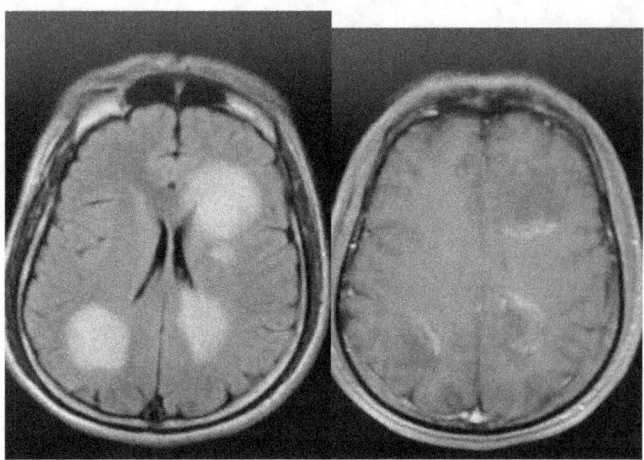

What is the diagnostic plan?
LP

If the above is negative, what should be done?
Stealth-guided biopsy for diagnosis

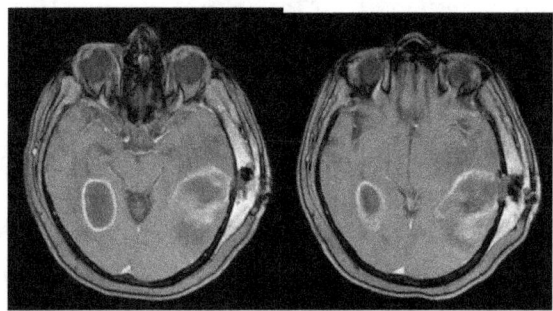

What is the most likely diagnosis in this patient with fevers, chills, elevated wbc count, and worsening altered mental status?
Multiple abscesses

What is treatment of choice in this patient who had prior left burr hole drainage of left temporo-occipital abscess, with significant worsening of mental status and increase in size of abscesses?
-OR for left craniotomy for abscess evacuation and right stealth guided burr hole drainage of abscess
-Abx per ID
-PICC line

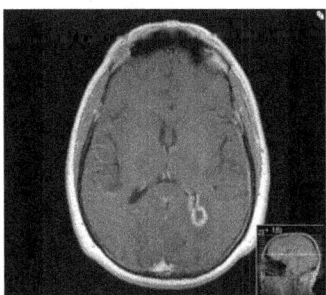

What is the differential diagnosis in this 56-year-old male patient presenting with acute onset severe headaches and neck pain?
Abscess, GBM, metastasis, tumefactive MS plaque

What is the most likely diagnosis in the above patient given diffusion B1000 MRI is positive and WBC are elevated with fevers and nuchal rigidity?
Abscess

What happened acutely to cause his severe headache?
Abscess ruptured into ventricle causing ventriculitis

What is best treatment plan?
-May do small volume LP to get organism and start empiric abx until Cx grows organism.
 -Need long term Abx
-May do stereotactic biopsy and drainage of abscess for diagnosis and empiric abx

What line must patient need for long term IV abx?
PICC—peripherally inserted central catheter

True or False. A central line and picc line have same degree of infection risk.
True

True or False. A central line or picc must be removed by 7 days or risk infection will increase dramatically.

False. Can both be kept in weeks or until patient has fevers

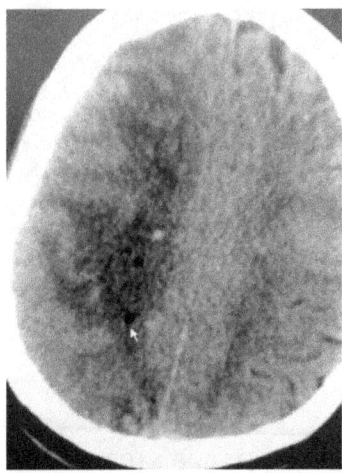

What is seen in this 67-year-old patient's CT head that is very concerning especially since he had never had any surgery?
Intracranial pneumocephalus within the parenchyma

What is this concerning for?
Intracranial infection
These spread quickly and patient may detiorate rapidly

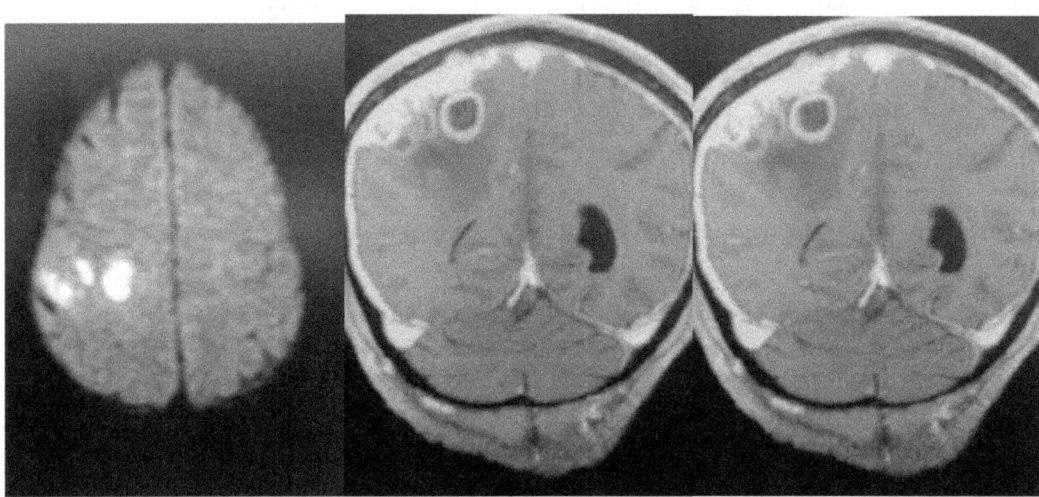

What is the differential diagnosis for the patient shown above?
Septic emboli, Abscess, Tumor with centers of necrosis

What is treatment of choice in this patient with AIDS and CD 4 count of 70 with wbc of 30,000 and CRP of 25?

-OR for stealth guided craniotomy for abscess evacuation—given that it is superficial can remove it entirely
-Then post-op abx long term

What service should be consulted post-op for treatment recommendations?
Infectious disease

What workup should be done in this patient?
HIV, CT Chest, panorex, ECHO heart

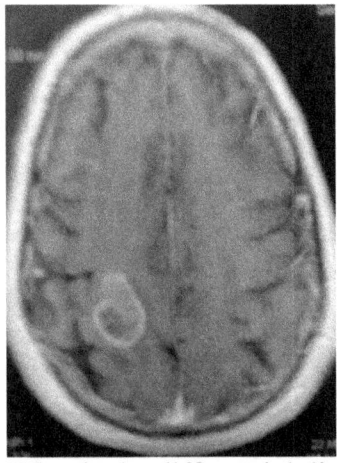

What is the differential diagnosis in the above 67-year-old patient who had an MRI of the brain due to tingling of left hand?
High grade glioma, metastasis, abscess, tumefactive MS plaque, Subacute stroke or hemorrhage

What is treatment recommendation for this patient with negative metastatic workup?

-OR for stealth-guided biopsy of this lesion
-May consider debulking with stealth guidance but careful for motor strip
- -Use fMRI pre-op to locate motor areas prior and intra-operative mapping as well.
Note: This was a high grade glioma and Gliadel was placed in cavity

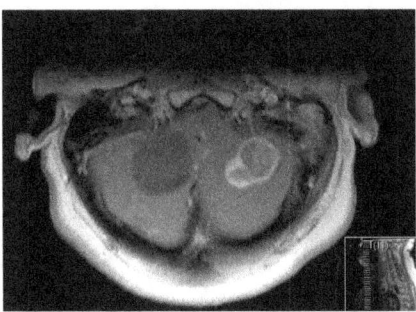

What is differential diagnosis for the lesion in the left cerebellar hemisphere in this 23-year-old female with ataxia?
Metastasis
Hemangioblastoma

Astrocytoma
Abscess—shows restricted diffusion on MRI due to pus in center
MS tumefactive plaque
GBM

What characteristic of a ring enhancing lesion distinguishes an abscess from another pathology most of the time?
Restricted diffusion in center
Capsule often incomplete on edge facing the ventricle due to lack of oxygen

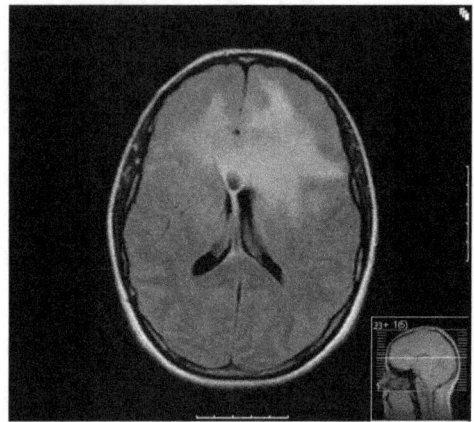

42-year-old female is brought in by family due to altered mental status. MRI is shown above. No contrast enhancement is noted contrasted studies.
What study is this?
FLAIR—fluid attenuated inversion recovery=T2 weighted with suppression of CSF

What is differential diagnosis?

GBM, lymphoma, encephalitis-west nile, other viral, Anaplastic astrocytoma

What is the term given to the pattern of edema signal that spread through the corpus callosum?
Butterfly glioma

What is diagnostic method of choice?
Low volume LP—send for cytology, west nile PCR.

An LP may be done if which are true?
Basilar cisterns are open, all ventricles communicate, mass effect is not severe

If those studies are not definitive, next choice is what?
Brain biopsy with Neuronavigation guidance

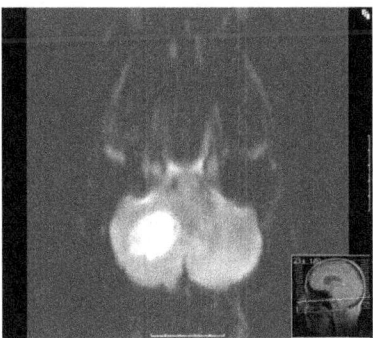

50-year-old male presents with altered mental status. He is very lethargic but still follows commands and opens his eyes to voice. He states his name and age. He has a fever of 40.1 C. His wbc count is 20. His CRP is elevated at 3. His ESR is elevated at 45. CT head showed large are of hypodensity in right cererebellar hemisphere with surrounding edema. MRI brain with and without contrast is shown above. What is the most likely diagnosis?
Cerebellar abscess

What MRI image is shown on the above picture?
B1000 diffusion image—shows large area of central restricted diffusion that matches on ADC map (apparent diffusion coefficient) consistent with pus.

What does the second MRI scan show?
Ring enhancing pattern

What is the next course of action after a full set of cultures are obtained from blood, sputum, etc?
-OR for Suboccipital Craniectomy and drainage of abscess
-If abscess was more superficial, then may resect the abscess and wall, but simple needle drainage through burr hole is sufficient with stealth guidance to collapse the capsule and allow antibiotic treatment

According too the operating room Time Out policy, antibiotics must be given pre-op. Is this the case in our patient here?
No. Hold antibiotics until samples taken then give antibiotics

According to new policy, what must be done pre-operatively to note the site of surgery on a patient?
Mark patient with indelible marker-not with erasable marker or sticker that may fall off.

What must be done post-operatively to find source of infection in our patient?
-CXR vs CT chest to r/o pulmonary nodules that may be seen in TB/Nocardia infections
-Panorex of teeth and physical examination of teeth to assess dentition
-TTE to r/o endocarditis

What consultation must be made post-operatively to aid in treating this patient?
ID consult

What long term central line must be placed for antibiotic therapy?
PICC line

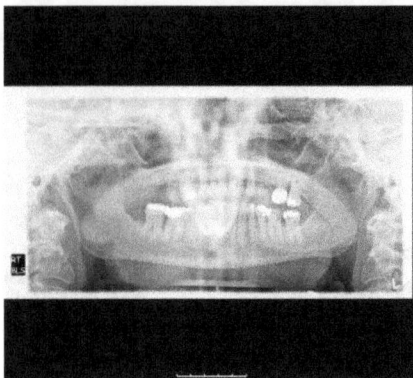

The above patient with cerebellar abscess has this study done per dental services consultation. What is this study and what does it show?
Panorex
Shows several missing teeth and treated cavities

Assuming the resident on call gets a call from the lab stating that the cultures have just grown out several partially acid fast bacilli(rods), Gram positive, catalase positive, strict aerobes. What is the most likely diagnosis?
Nocardia

What is treatment of choice?
TMP-SMZ (Bactrim)

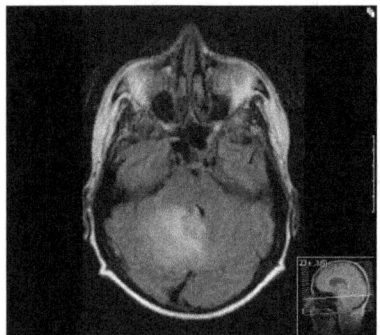

What does this post-op MRI on the above patient demonstrate?
Edema within drainage bed of abscess and still distortion of 4th ventricle

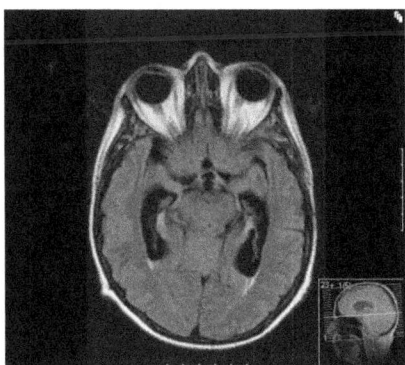

What does this MRI scan show that is worrisome for continued increased ICP?
Large temporal horns

According to the literature, what situation may portend a 50% mortality in a patient with cerebral abscess?
Rupture into the ventricles

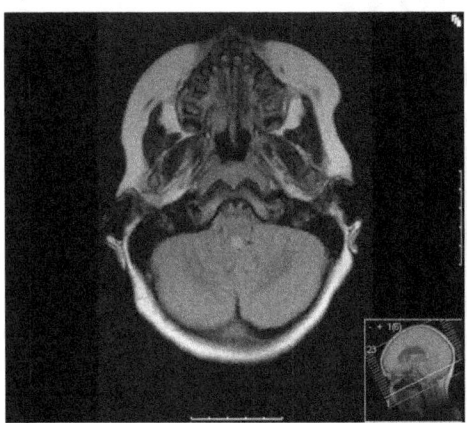

15-year-old female from Mexico presents with headache. She is neurologically intact. MRI is shown above. What is the most likely diagnosis?
Neurocysticercosis at outlet of 4th ventricle/foramen of magendie

What is the next best course of action according to the literature?

-May treat this with albendazole and steroids.
-Avoid OR emergently unless change in neuro exam as can cause spread of infection through csf.
-Note: Some have treated this endoscopically through right frontal burr hole

What is classically seen in a patient with neurocysticercosis on CT scan or MRI?
Multiple lesions in different stages

Calcified lesions are active lesions?
False

Lesions with edema around them are active lesions?
True

What are the two types of neurocysticercosis forms?
Racemose and cellulose

What serum marker can be sent in a patient with cysticercosis?
Antibody to neurocysticercosis

The foramina of Luschka are located medial or lateral to the the fourth ventricle?
Lateral

Foramen of Magendie is located medial or lateral?
Medial

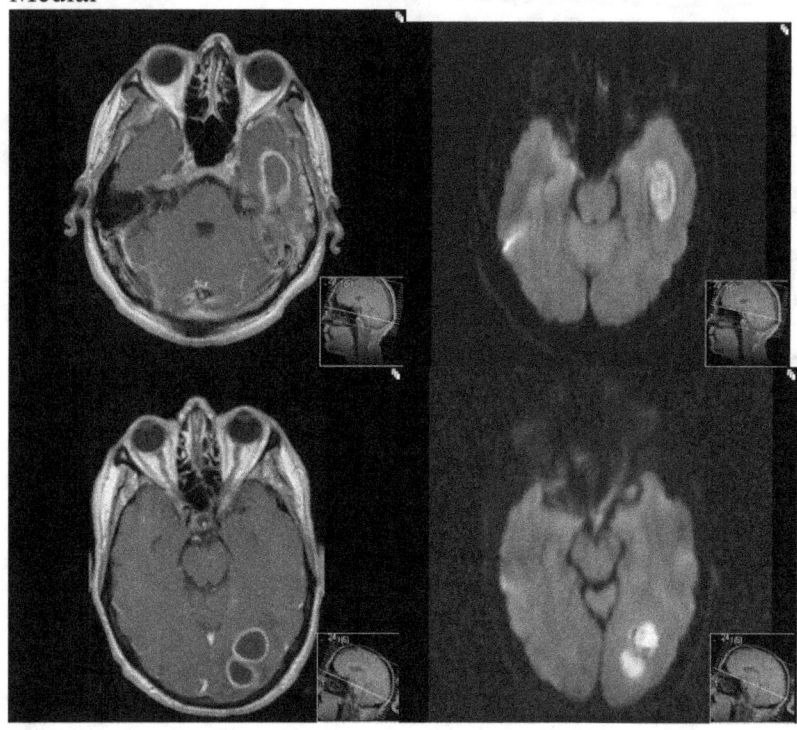

What is the most likely diagnosis for the two above patients?
Abscesses

What is the best treatment for the temporal abscess?
Needle drainage using stealth guidance and abx

What is the differential for the temporal lesion?
Abscess, GBM, metastasis, tumefactive MS, subacute stroke

Why is an abscess most likely?
Discharge from ear, proximity to internal auditory meatus

What is the best treatment for the occipital abscess?

Needle drainage and abx using stealth guidance

True or False. If an abscess comes to the surface of the brain, it should not be removed via craniotomy
False. Remove it with craniotomy since so superficial

True or False. If an abscess is deep in the brain, it should be treated via craniotomy only.
False. If deep, needle drain it and can obtain organism.

True or False. If a patient has an abscess than is small and is neurologically stable, and has blood cultures positive for staph aureus that is methicillin-resistant, then it is fine to do abx therapy and repeat scans to see if abscess has resolved, rather than go straight to surgery.
True. But if increase in size, might question polymicrobial nature of abscess that would require a change in abx therapy vs needle biopsy abscess.

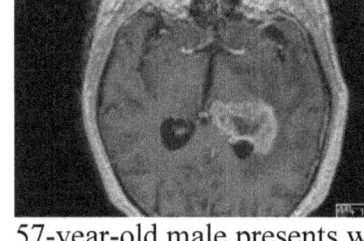

57-year-old male presents with double vision and right sided weakness and headache. MRI is shown above. Describe the findings in the above MRI with and without contrast?
Heterogeneously ring-enhancing mass 3x4 cm in size within posterior thalamus left side abutting the atrium of left lateral ventricle with surrounding vasogenic edema

What is differential diagnosis for this lesion?
GBM
Anaplastic astrocytoma
Abscess/infection
MS tumefactive plaque
Lymphoma

What must not be given to a patient suspected of having a lymphoma that requires biopsy?
Steroids as they can causes miraculous disappearance of lymphomas (ghost lesion)

The above patient had a neuronavigation guided biopsy. The path showed necrosis, mitotic figures, and vascular and endothelial proliferation. What is the diagnosis?
GBM

What is the next course of action?
Oncology consultation for chemo and radiation

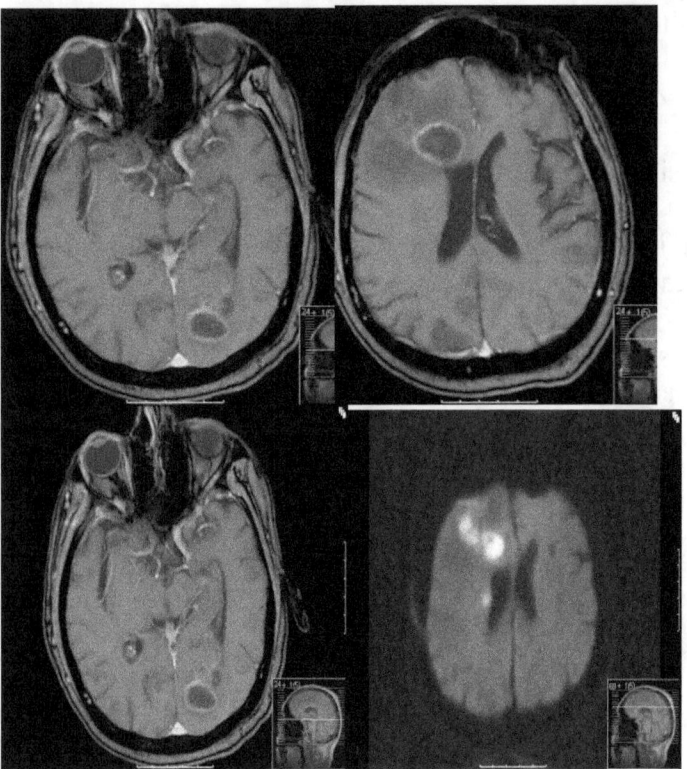

How might the above patient with the MRI shown above present?
Fevers, chills, headaches, double vision

What studies are shown above?
MRI brain with contrast

What is the differential diagnosis for a patient with the above findings?
Abscess, Multifocal GBM, Metastasis, tumefactive MS lesions, multiple subacute strokes, lymphoma

What study would one like to see to confirm our diagnosis of abscess?
Diffusion b1000-hyperintense in center due to restricted diffusion from pus

What is seen on these MRI images that produces a more severe prognosis for this patient?
Rupture into right lateral ventricle with pus in atrium

What is best course of action?
Stealth guided drainage of right frontal abscess and Abx

What feature of abscess wall causes increase risk of rupture into ventricle?
Thinner wall on side facing ventricle

GBM capsules are different than abscess capsules as they are:
Much thicker

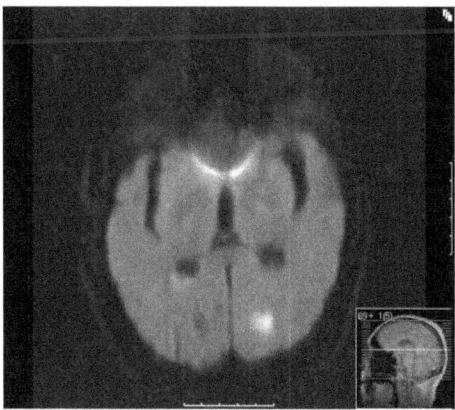

What is noted on this diffusion MRI scan portending a worse prognosis for the above patient?
Pus in right atrium lights up, also ventriculitis picture if inner wall of entire ventricular system + on diffusion

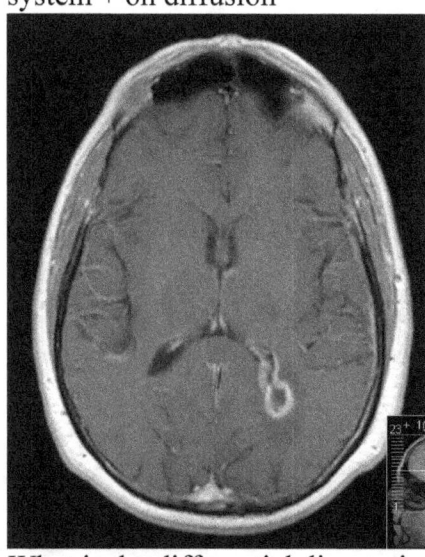

What is the differential diagnosis in the 56-year-old male patient presenting with acute onset severe headaches and neck pain?
Abscess, GBM, metastasis, tumefactive MS plaque

What is the most likely diagnosis in the above patient given diffusion is positive and WBC elevated with fevers and nuchal rigidity?

Abscess

What happened acutely to cause his severe headache?
Abscess ruptured into ventricle causing ventriculitis

What is best treatment plan?
-May do small volume LP to get organism and start empiric abx until Cx grows organism.
-Need long term Abx
-May do stereotactic biopsy and drainage of abscess for diagnosis and empiric abx

What line must patient need for long term IV abx?
Picc—peripherally inserted central catheter

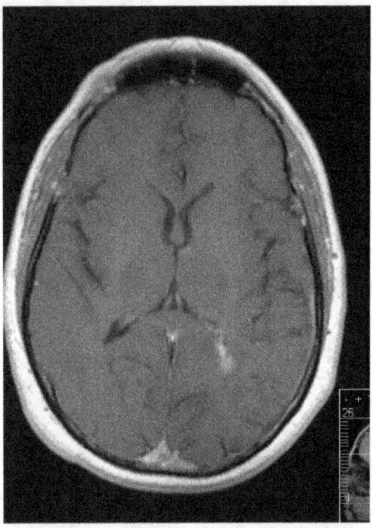

Repeat MRI Brain with and without contrast done 4 weeks later showing what?
Decrease in size of abscess—abx are working

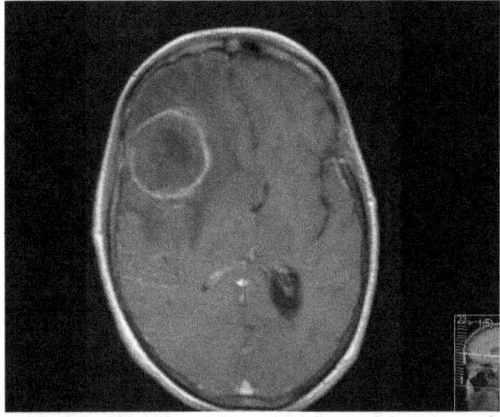

50-year-old female with long history of smoking presents with a seizure. MRI is shown above with contrast. Describe the above finding?
Right frontal ring enhancing lesion with central hypointensity with surrounding vasogenic edema and right to left midline shift

What is differential diagnosis?
Metastasis, GBM, abscess, tumefactive MS,

What is next study to work up this patient?
CT chest/abd/pelvis

CT C/A/P shows a large left lung mass. Oncology states that she has longer than 6 months to live. What is next treatment strategy?

-Given midline shift and seizures, recommend surgical resection with stealth guidance. Thus can obtain diagnosis and decrease mass effect. No other lesion noted on MRI
-Gamma knife can be used but usually for lesions smaller than 3 cm
-Start on Seizure medication

What other treatments will be needed post resection?
Whole brain radiation therapy
Chemotherapy

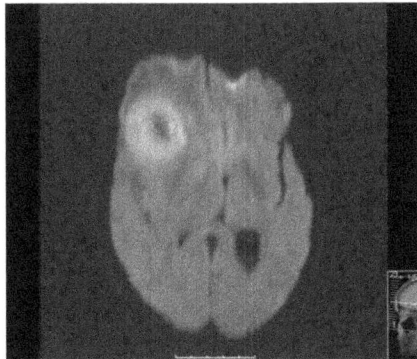
The above study in this patient confirms what finding?
This is not an abscess

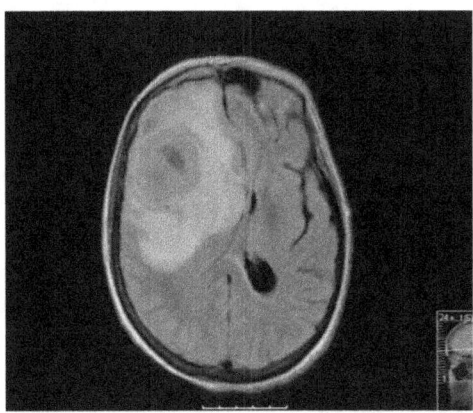
What study sequence is the above and what does it show?
FLAIR, lots vasogenic edema

Vasogenic edema is formed from what?
Breakdown of blood brain barrier

What medication helps decrease this edema?
Decadron-steroids

Patients on long term steroids who present after big health problem to hospital can become hypotensive and hyponatremic. What should be given to these patients?
Stress dose steroids—hydrocortisone 100 TID

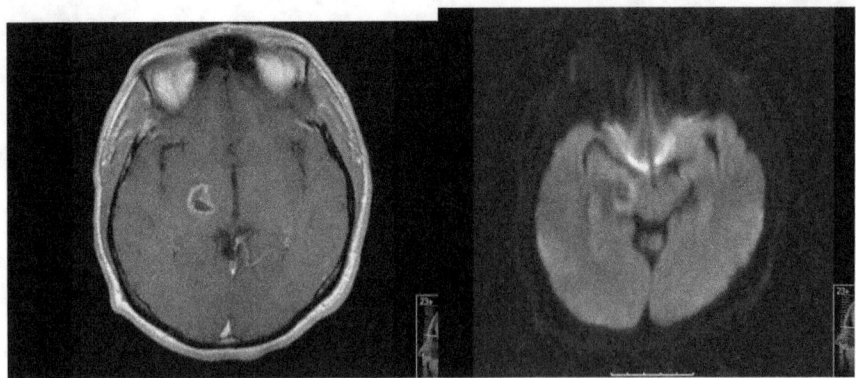

What other study should be done next for diagnosis?
CT C/A/P

What is the differential diagnosis?
Primary tumor like GBM, abscess, MS plaque, subacute stroke, oligo

CT C/A/P shows a large midline lung mass. What is next step in diagnosis?
IR guided bronchoscopic biopsy of lung mass for pathology.

CT C/A/P is negative. What is the next step in diagnosis?
Stealth-guided biopsy of mass in Brain

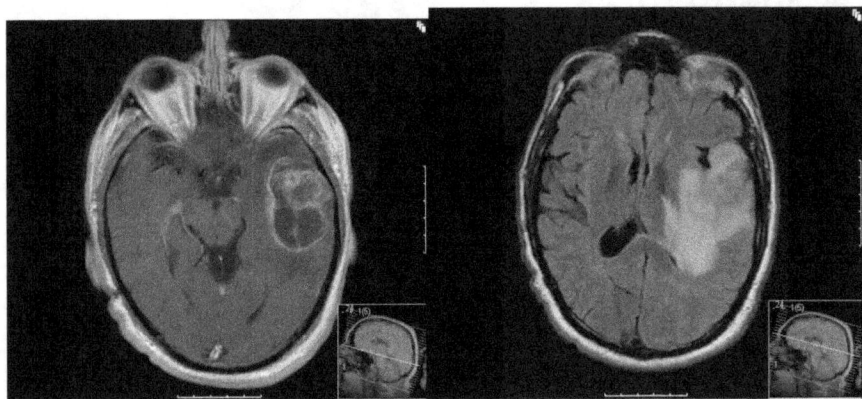

What is the differential diagnosis?
GBM, abscess, tumefactive MS plaque, traumatic contusion, large subacute infarct

What is best treatment strategy?
OR for craniotomy to debulk mass—as it is causing mass effect

The second MRI above is what sequence and what does it show?
FLAIR, vasogenic edema

True or False. Intra-operatively GBM tissue can look exactly like normal brain without any clear plane.
True

The above tumor was a GBM. Gliadel was placed into the cavity after safe debulking based on the frozen specimen diagnosis intra-operatively. What is postoperative treatment regimen?
Chemo (temozolamide) and XRT

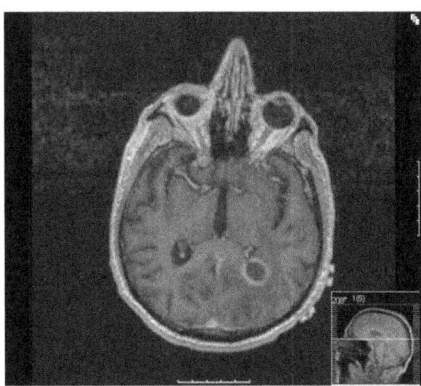

What is differential diagnosis for the above patient with bad dental hygiene?
Abscess, primary tumor GBM, metastasis, MS plaque

What is dangerous about this particular lesion that can increase mortality dramatically?
Adjacent to ventricle, can blow into ventricle and cause ventriculitis

What is best treatment strategy if patient presented with severe headaches acutely?
-Likely blew into ventricle, thus do LP small volume and get culture, then start abx long term
-If no acute history that blew into ventricle, do stealth guided aspiration then start abx

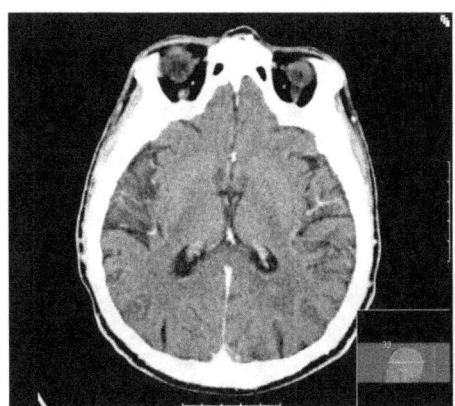

3 months later a CT head is obtained. What is seen on this scan?
Abscess gone 3 months later after long term abx

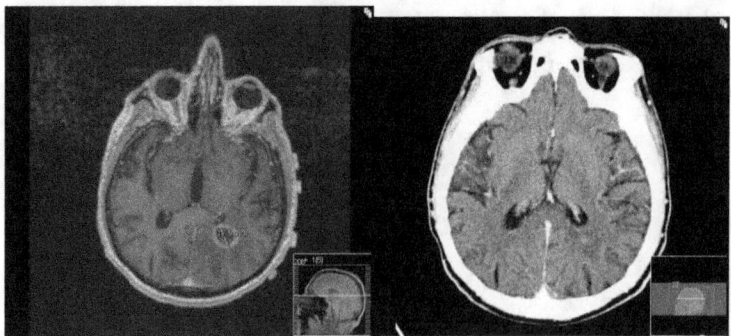

What is the differential diagnosis in the above patient?
Primary tumor GBM, metastasis, Abscess, MS plaque, subacute infarct or hemorrhage

What is treatment of choice?
Stealth guided Needle drainage and culture of sample

Who should be consulted post-op if sample returns with pus?
ID

What workup should be done to find source of abscess?
CXR, CT chest look for nocardia nodules, panorex of teeth and dental consult, blood Cx, TTE of heart to look for endocarditis

Culture grew out strep milieri. What is likely source?
Teeth

Cx grew out nocardia. What is likely source?
Lungs

What is treatment for nocardia?
Bactrim 6 weeks

When should another MRI head be done?
After complete abx or if change neuro status

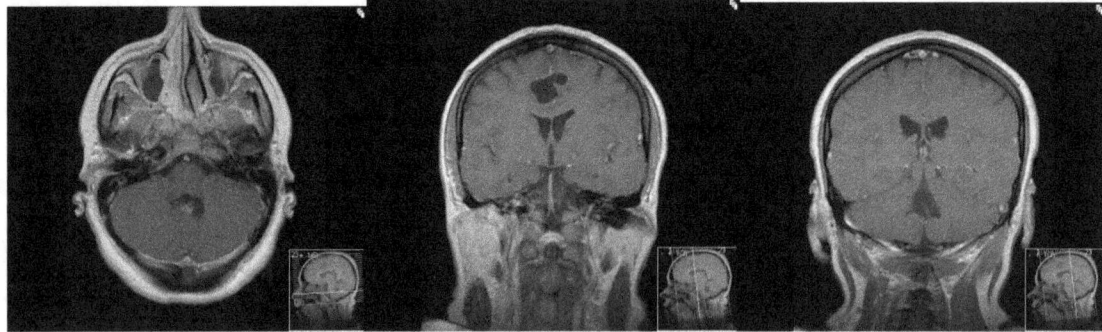

What is the likely diagnosis?
Neurocysticercosis

Where is the cyst in the ventricular system in the above patient?

4th ventricle

What would the CT head show?
Areas of edema around active cysts and calcifications around inactive cysts

Patient has headaches but is neuro intact. No papilledema noted. Venous pulsations in the eyes are present. What is treatment?
Antihelminthics such as albendazole with steroids

Should the cyst in 4th ventricle be taken out in surgery immediately?
No, albendazole can treat these according to literature
Surgery has risk of seeding cysts and causing communicating hydrocephalus

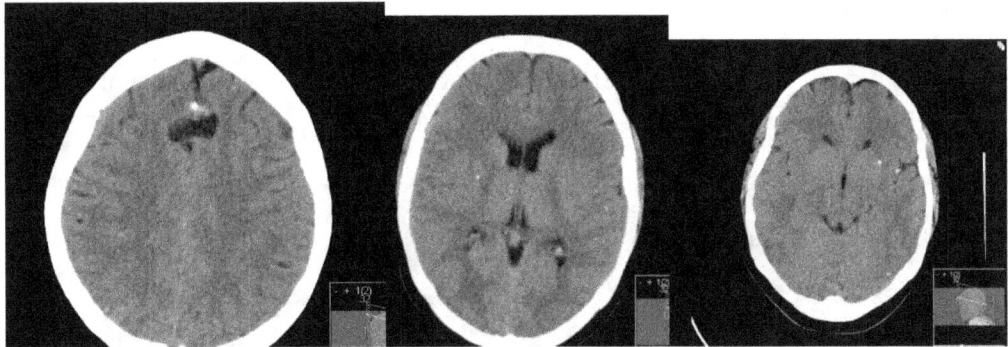

What is the diagnosis based on the above CT scans?
Neurocysticercosis

Which stage is associated with a silent dormant cyst?
Calcified

True or False. Patients with this pathology can present with seizures and/or hydrocephalus depending on the location of the cysts.
True

What is the next step for the above patient with seizures?
Seizure control per neurology, send serum cysticercosis labs, MRI brain with contrast, steroids and praziquantil/albendazole

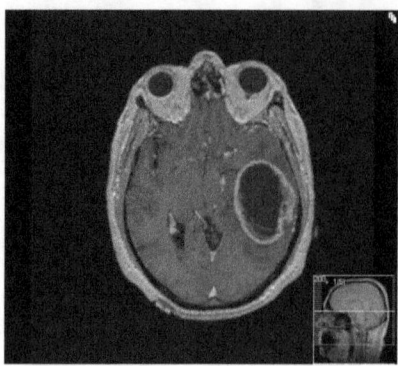

What is differential diagnosis in this 67-year-old male?
GBM, abscess

What surgery can be done for this cystic tumor?
Ommaya reservoir into tumor

True or False. The patient should be on seizure medication even though he has not had seizures
False

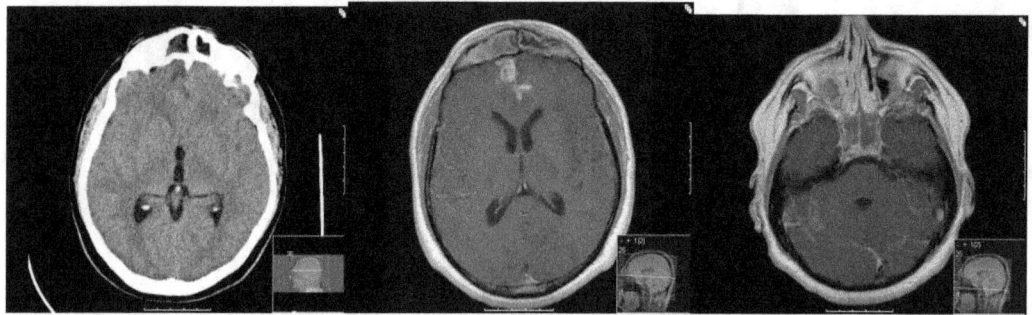

45-year-old male s/p frontal sinus surgery 2 weeks ago with fevers and chills, and wbc increase. What is most likely diagnosis?
Late cerebritis frontal lobes involving frontal sinus and ethmoids which are packed with pus—not yet early abscess formation

What is treatment?
Consult ENT for washout of sinus, get Cx, Abx

Does this patient need seizure medication?
Yes 7 days

What tumor can cause weakness of both legs acutely without any other findings?
Bifrontal meningioma

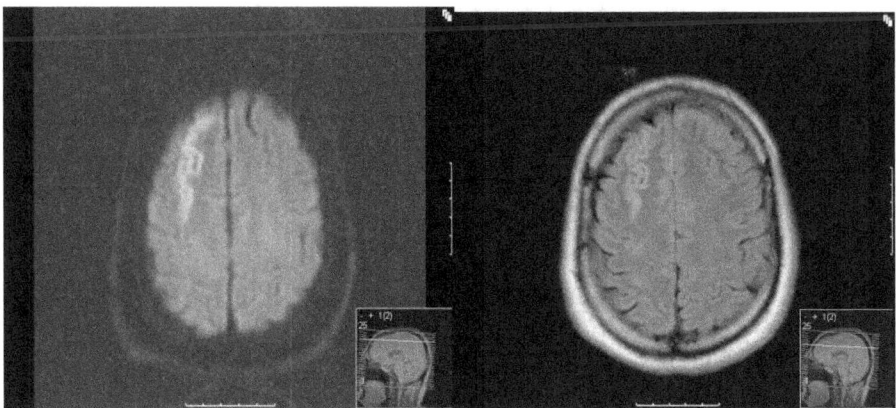

What is seen on the diffusion MRI in the above patient?
Cerebritis-Pus

True or False. FLAIR edema is commonly seen in cerebritis infections
False. No edema with infections. Only with strokes and tumors

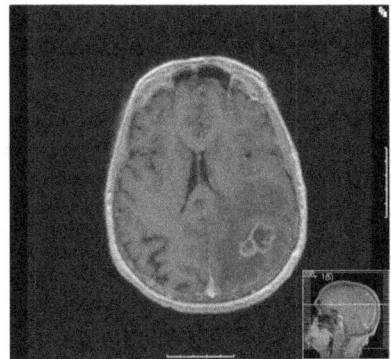

What is the differential?
GBM, abscess, Met

What other imaging study should be done?
CT c/a/p and CXR

Assuming there is a large lung mass, and there are no other lesions, what treatment should be offered to this otherwise healthy patient with high karnofsky score?
Gamma knife (less than 3 cm)

What lesion should be biopsied?
Lung mass to obtain diagnosis

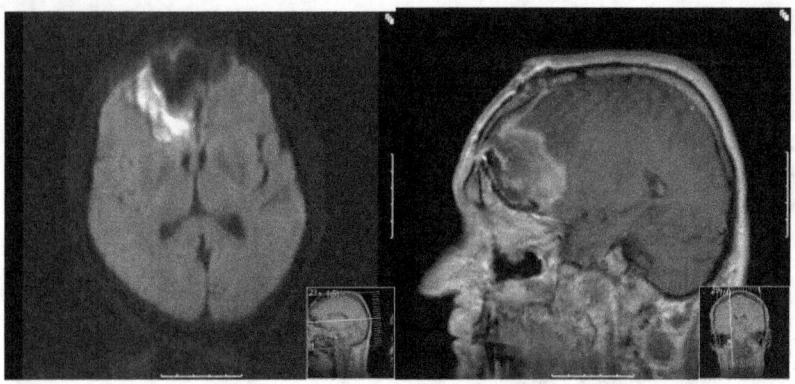

What is diagnosis?
Frontal abscess due to frontal sinus fracture that was untreated

What is treatment?
OR for bifrontal craniotomy for abscess evacuation
Do with ENT

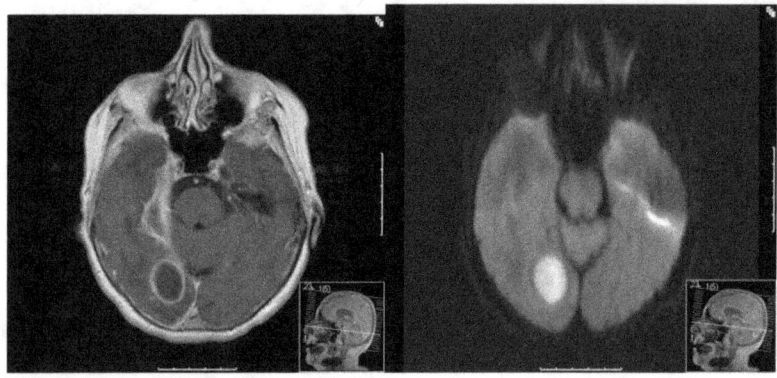

What is the likely diagnosis?
Right abscess

What is treatment?
Stealth guided drainage
Blood Cx
Abx

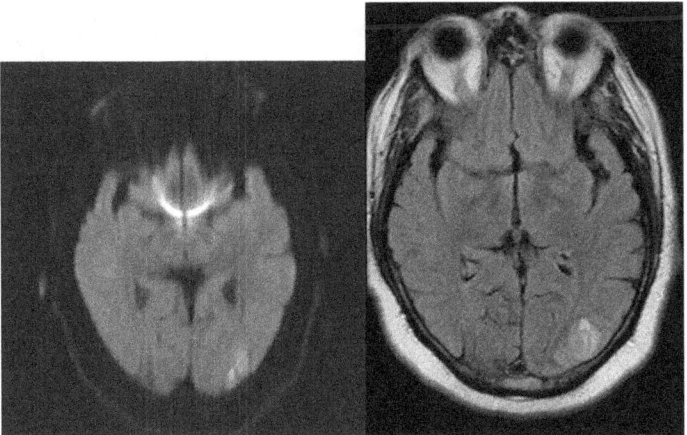

What is the diagnosis in this patient with fevers and headaches and right visual field scotomas?
Left occipital lobe Abscess

What MRI study above helps to diagnosis an abscess?
+ restricted diffusion on B 1000

What is the 2nd MRI study shown above?
FLAIR

Is there significant surrounding edema around the abscess?
No

What Neuro medications should the patient be placed on when admitted to the ICU?
Steroids, Anti-Epileptic medication

What treatment is recommended?
Steroids-especially if lots of edema, studies showing better results and no increase risk of abscess growing
Antibiotics-empiric first, then specific antibiotics once cultures come back with sensitivities

What labs must be drawn to obtain organism?
Pan Cx of blood, urine, sputum
CXR
CT chest r/o nocardia nodules
Panorex of teeth
TTE r/o endocarditis

What consultations should be made?
Dental
ID

What are some risk factors for intracranial abscess?

IVDA, Diabetes, immunosuppression drugs like steroids long-term for RA patients, chemotherapy patients

What would be the recommendation for a large deep abscess causing mass effect?
Stealth guided drainage

What about for large superficial abscess causing mass effect?
Craniotomy for resection

What about multiple small deep and superficial abscesses without mass effect?
-Abx and serial MRI to see if increase or decrease in size
-The above patient has small abscess with no mass effect, abx should work well

When would an MRI with contrast be repeated?
After abx treatment completed

Cultures came back as G+ rods. CT chest shows nodules. What is likely diagnosis?
nocardia (G+ rods)

What is treatment?
Bactrim

Bactrim is a member of what family of antibiotics?
Sulfa drugs

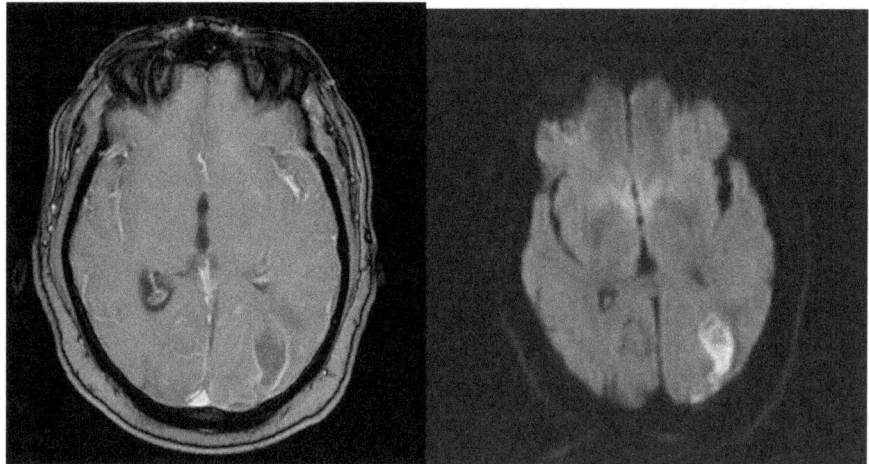

What is the differential diagnosis in this 54-year-old male?
Abscess, GBM, metastasis, MS plaque, subacute stroke or hemorrhage,

This patient has fevers and chills, elevated ESR and wbc count. What is likely diagnosis?
Abscess

Where is the abscess?

Left occipital lobe

What is the second MRI scan shown above?
MRI brain with contrast-showing ring enhancing mass

What is noted about the capsule of an abscess that is different from a tumor?
Usually thinner at ventricular end –may rupture into ventricle
Capsule usually thinner throughout than the capsule of Tumor like GBM

What is treatment for this patient with large abscess in occipital lobe with mass effect?
Left occipital craniotomy for resection of abscess

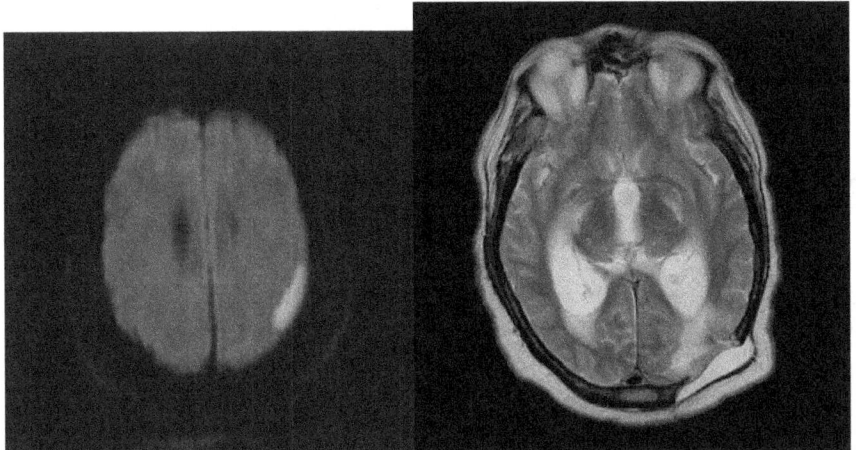

What is seen on this T2 MRI that is concerning with a patient with detiorating neuro exam?
Triventricular enlargement from obstruction from abscess debris floating into the ventricle obstructing the arachnoid villi causing a communicating hydrocephalus

What treatments can be offered?
Endoscopic Third Ventriculostomy (ETV) vs VPS (Once infection treated)
EVD drainage in the meantime and follow CSF cultures

What area of restricted diffusion is noted in this patient?
Extra-axial collection of likely pus

What is recommended treatment?
OR for evacuation of subgaleal empyema versus subdural empyema

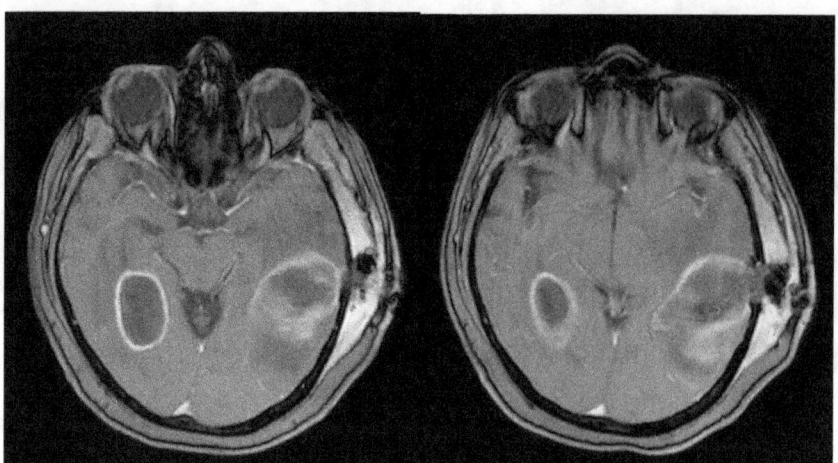

What is the most likely diagnosis in this patient with fevers, chills, elevated wbc count, and worsening altered mental status?
Multiple abscesses

What is treatment of choice in this patient who had prior left burr hole drainage of left temporo-occipital abscess, with significant worsening of mental status and increase in size of abscesses?
-OR for left craniotomy for abscess evacuation and right stealth guided burr hole drainage of abscess
-Abx per ID
-PICC line

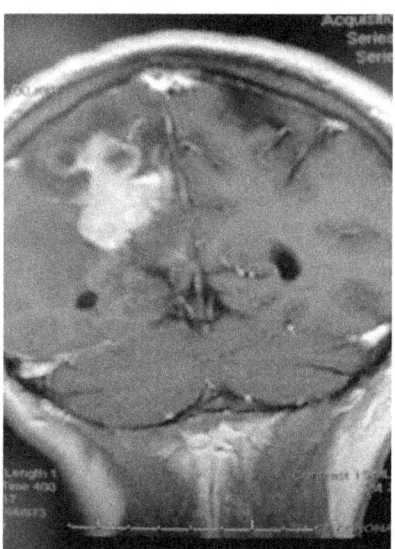

What is the likely diagnosis in this HIV-positive patient with fevers and chills?
intracranial abscess

What is the treatment?

Stealth-guided Craniectomy and abscess resection

True or False. After removing the abscess, the bone flap can be place back on with screws and plates.
False. Do not place back on. It is infected.

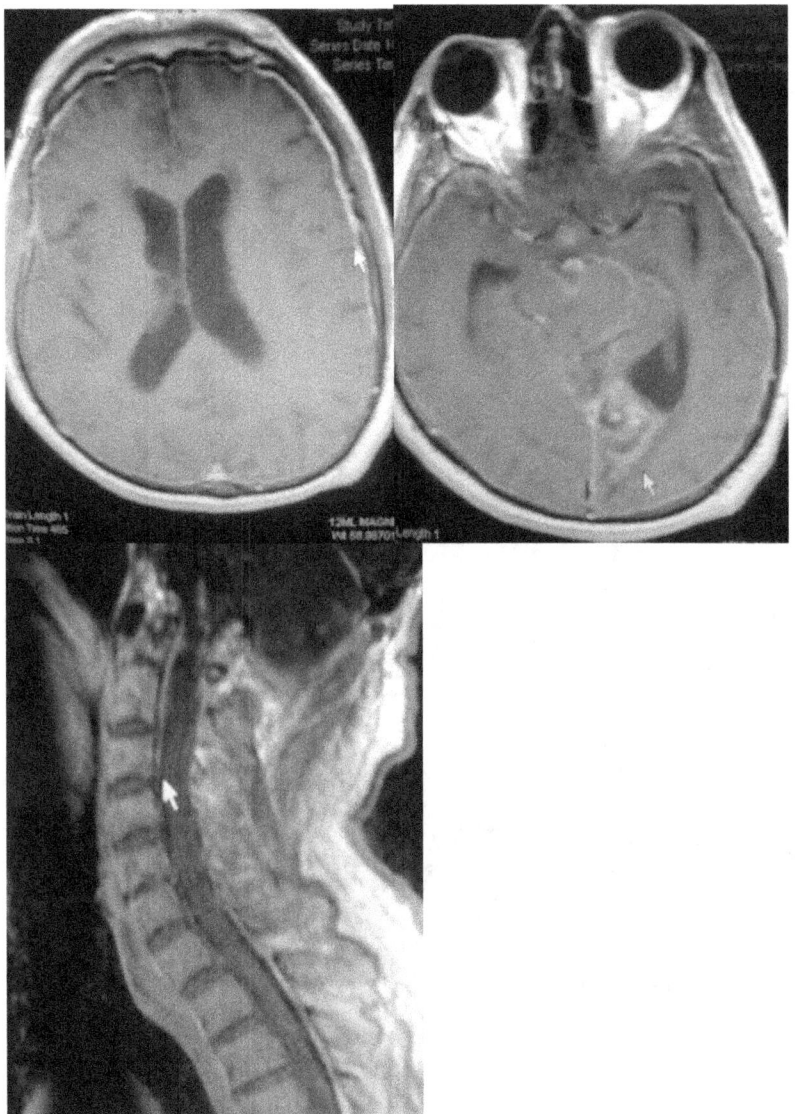

What are the arrows pointing at in this 68-year-old male with altered mental status?
Likely leptomeningeal enhancement diffuse

What are some causes for leptomeningeal enhancement?
Lymphoma, carcinomatosis, post-large volume LP or CSF Leak, meningitis

What fluid should be analyzed?
CSF

True or False. Epidural or subdural collections of pus can be seen in the spine as well.

True

True or False. Epidural abscesses of the spine are considered emergencies if patient has acute neurological symptoms such as cauda equina.
True

What surgical treatment may be offered for spinal epidural abscesses?
Laminectomy for drainage of epidural collection

True or False. A drain is recommended post-operatively after removing spinal epidural abscess.
True

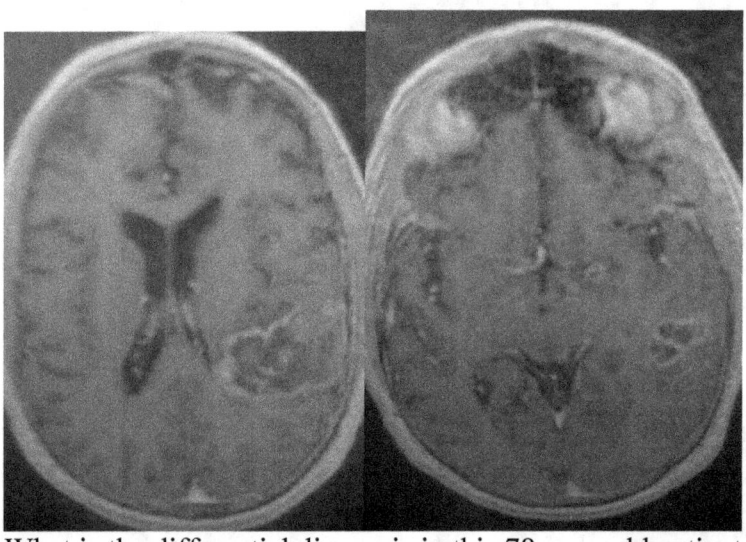

What is the differential diagnosis in this 78-year-old patient with headaches and right sided weakness?
GBM, Tumefactive MS Plaque, Abscess, Subacute stroke or hemorrhage

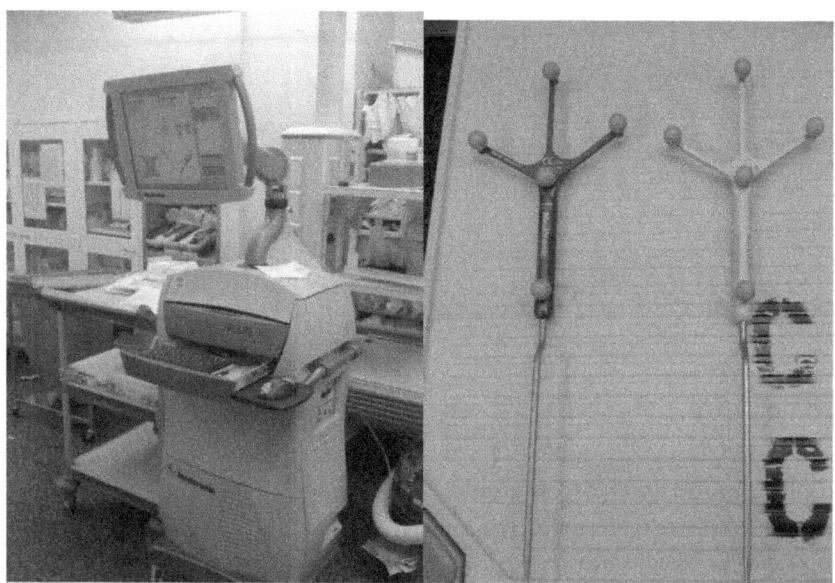

What is this technology?
Stealth neuronavigation

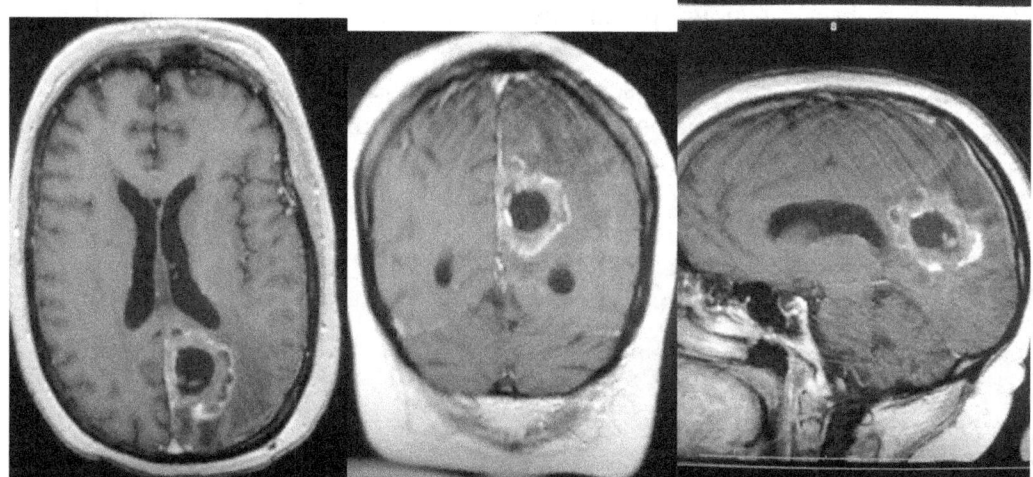

What is the differential diagnosis in this patient with Metastatic breast cancer?
Metastasis

What is treatment for this patient with 2 year life expectancy?
OR for craniotomy and tumor resection

What adjuvant treatment will be needed post-op?
Chemo and radiation per oncology

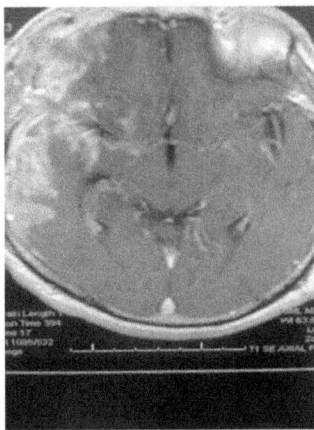

This 56-year-old male is s/p clipping of aneurysm now with fevers and chills. MRI shown above. What is likely diagnosis and treatment?
-Cerebritis, OR for washout and bone removal—remove all foreign body and leave bone out, do not place anything in other than a drain
-Post-op ABX and picc line, FU cultures, ID consult

A patient with brain abscess is now s/p drainage. What should be done with bone flap intra-operatively?
-If no subdural empyema or epidural abscess, can place bone back on
-If empyema that has devascularized bone, then leave bone out

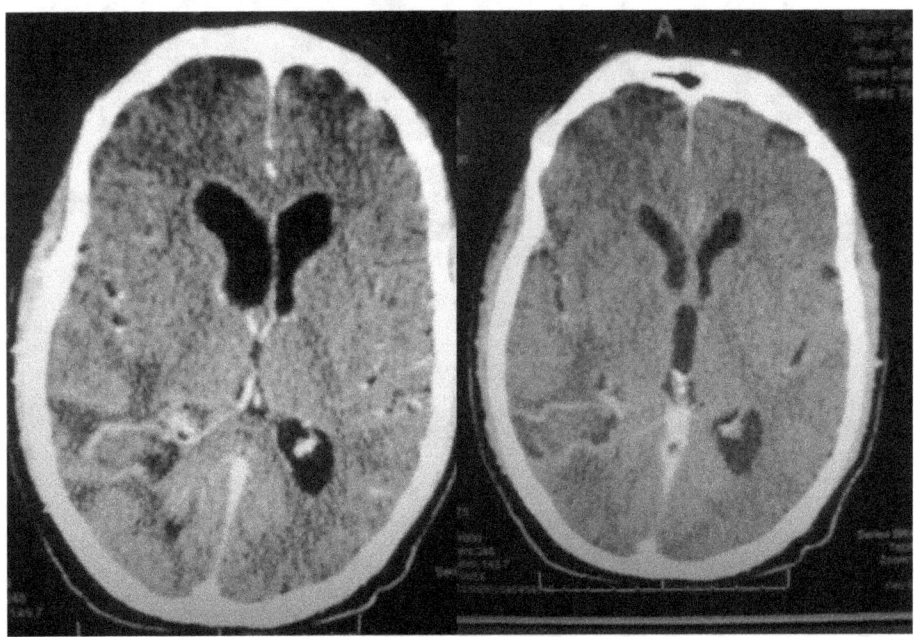

What is the most likely diagnosis in this febrile 34-year-old female with poor dentition and elevated wbc count?
Abscess

What factor in the above scan points to a high mortality in this patient?

Rupture into the ventricle

What is the treatment?
EVD placement and aspiration of pus and intrathecal gentamicin and ceftriaxone

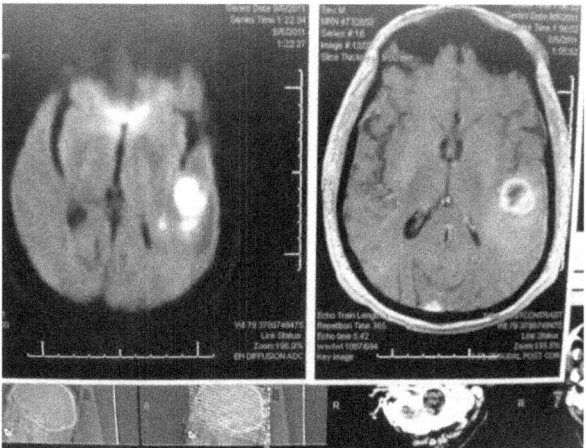

45-year-old diabetic male presents with headaches and difficulty with speech comprehension and repetition. He has a fever of 38.6 and wbc count of 16. He is missing some teeth as well. What type of aphasia causes repetition difficulty?
Conduction aphasia

Between what two areas does the above aphasia localize to?

The fibers connecting broca's areas in inferior frontal lobe with wernicke's area in posterior superior temporal gyrus

What is the differential diagnosis in the above patient with the MRI with contrast shown?
Abscess, Tumor (Low or high grade astrocytoma), tumefactive MS plaque, Subacute stroke or hemorrhage

What is the most likely diagnosis based on the MRI scans shown above?
Abscess—DWI B1000 + indicating central area of pus

What is recommended treatment?
OR for axiom (stealth-guided) drainage of abscess, then start abx after drain it

What further workup is needed in this patient post-operatively?
Panorex, dental consult and possible removal of infected teeth, TTE to look for infective endocarditis, picc line and ID consultation for abx

What other sources of infection could cause the above besides teeth?
Fungal (Nocardia) from inhaling spores

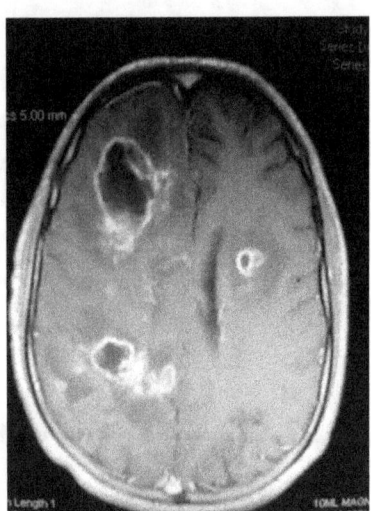

What is the most likely diagnosis?
Intracranial abscesses

What is the treatment?
Stealth guided drainage of large abscess right frontal, then abx therapy per ID

SUMMARY:

In this textbook, we have described many cases of diagnosis and management of neurological infectious disease states. We include many pictures of MRI and CT scans to aid in the teaching and learning for each of these case scenarios. We hope that readers can use this information in their future interactions with their own patients as they progress through their medical careers. We wish you continued success and strength in caring for the sick patients that will undoubtedly come to your door.

The Authors

Hippocratic Oath

I swear by all that I hold most sacred
That I will keep this enduring oath:

To the best of my ability and judgement I will
Practice the Art only for the benefit of my patients.

Whatever houses I may visit, I will enter only
To help the sick or to prevent illness,
Never to inflict harm, injustice, or suffering.

I will lead my life and practice the Art
Conscientiously and with honor.

Whatever I may see or hear in the practice of the Art
Or even outside of it that should not be spread abroad
I will keep in solemn confidence.

I will be just and generous to those who taught me
The Art, to my colleagues in the Art, and to
Those who desire to learn it.

May happiness and the physician's good repute be
Granted me while I keep this sacred Oath inviolate.

www.ingramcontent.com/pod-product-compliance
Lightning Source LLC
Chambersburg PA
CBHW080854170526
45158CB00009B/2730